KT-131-784

THE NATIONAL TRUST
Little Library

Exotic FRUITS

JILL NORMAN

DORLING KINDERSLEY
LONDON • NEW YORK • STUTTGART

A DORLING KINDERSLEY BOOK

EDITOR MARK RONAN

SENIOR EDITOR CAROLYN RYDEN

DESIGN MATHEWSON BULL

PHOTOGRAPHER DAVE KING

FIRST PUBLISHED IN GREAT BRITAIN IN 1992
BY DORLING KINDERSLEY LIMITED
9 HENRIETTA STREET, LONDON WC2E 8PS

A CIP CATALOGUE RECORD FOR THIS BOOK IS AVAILABLE
FROM THE BRITISH LIBRARY.

ISBN 0-86318-789-7

PRINTED AND BOUND IN HONG KONG
BY IMAGO

CONTENTS

INTRODUCTION

THE LAST DECADE *has seen a great increase in the availability of unusual fruits, many of them from the tropics. No longer restricted to specialist shops, they are widely sold in supermarkets. The musky perfume, with a hint of fermentation behind it, that is characteristic of many ripe tropical fruits can now be savoured without the need for long-distance travel. If you are susceptible to the sweet aromas of the tropics, try the lime, honey and jasmine fragrance of passion fruit; its intensity is not dissipated by freezing or by combining it with other ingredients, and passion fruit juice, with its light acidity, is an excellent alternative to orange juice for breakfast. Ripe feijoas and guavas have similar penetrating aromas, combining those of pineapple, gooseberry and lemon.*

More subtle and complex perfumes and flavours come from mangoes and papayas. Really ripe fruits of good quality are hard to find, with their aromas of peach and apricot, melon and spice, but even poorer specimens can make a useful contribution in the kitchen. Some of the tropical fruits we buy lack the lusciousness they would have when picked ripe. Some are bland, others too acidic to eat raw and dull when cooked. Mangosteens, for instance, taste superb when picked ripe and eaten at once, but unfortunately what reaches the shops is rarely of such quality.

The influx of tropical fruits owes much to the determined efforts of farmers and marketing men as far apart as Florida and Kenya, New Zealand and Colombia. After the success of the kiwi fruit with nouvelle cuisine *chefs, and its acceptance at home, the New Zealanders in particular have put great energy into growing cherimoyas, passion fruits, physalis, guavas and tamarillos.*

Clever breeding has led to the development of the sharon fruit, a handsome but bland version of the persimmon, which can be picked, stored and eaten when barely ripe. The persimmon, on the other hand,

Quince merchants,
Istanbul, c.1890

soft and blotchy, may not look up to much, but tastes wonderful, as generations of Japanese poets have testified.

Many fruits that flourish in a Mediterranean climate are becoming more popular: loquats and figs, pomegranates and prickly pears. The quince will ripen even in temperate climates, and its penetrating aroma, part apple, part pear, part pineapple, will fill the kitchen. It has not yet become widely marketed, but can be hunted down in Greek and Middle Eastern shops or begged from a friend with a tree.

CARAMBOLA, FEIJOA & GUAVA

NATIVE TO JAVA, *the carambola or star fruit,* Averrhoea carambola, *grows throughout the tropics, producing three crops a year of striking golden fruits.*

The waxy, thin skin and crisp flesh are both eaten. Ripe fruit smells slightly of pineapple and tastes pleasantly acid; unripe it is best cooked.

Carambola section

Prized as an ornamental evergreen with large red flowers, feijoa, *Feijoa sellowiana,* belongs to the myrtle family. The egg-shaped fruit has a gritty texture like some pears and is slightly tart. Ripe feijoa has an intense aroma of lime, gooseberry and pineapple.

Sliced feijoa

Also of the myrtle family, guava, *Psidium guyava,* is native to Brazil but is grown also in Asia. The round or pear-shaped fruit may have cream or pink flesh. It can smell gently musty or strong and penetrating.

Feijoa

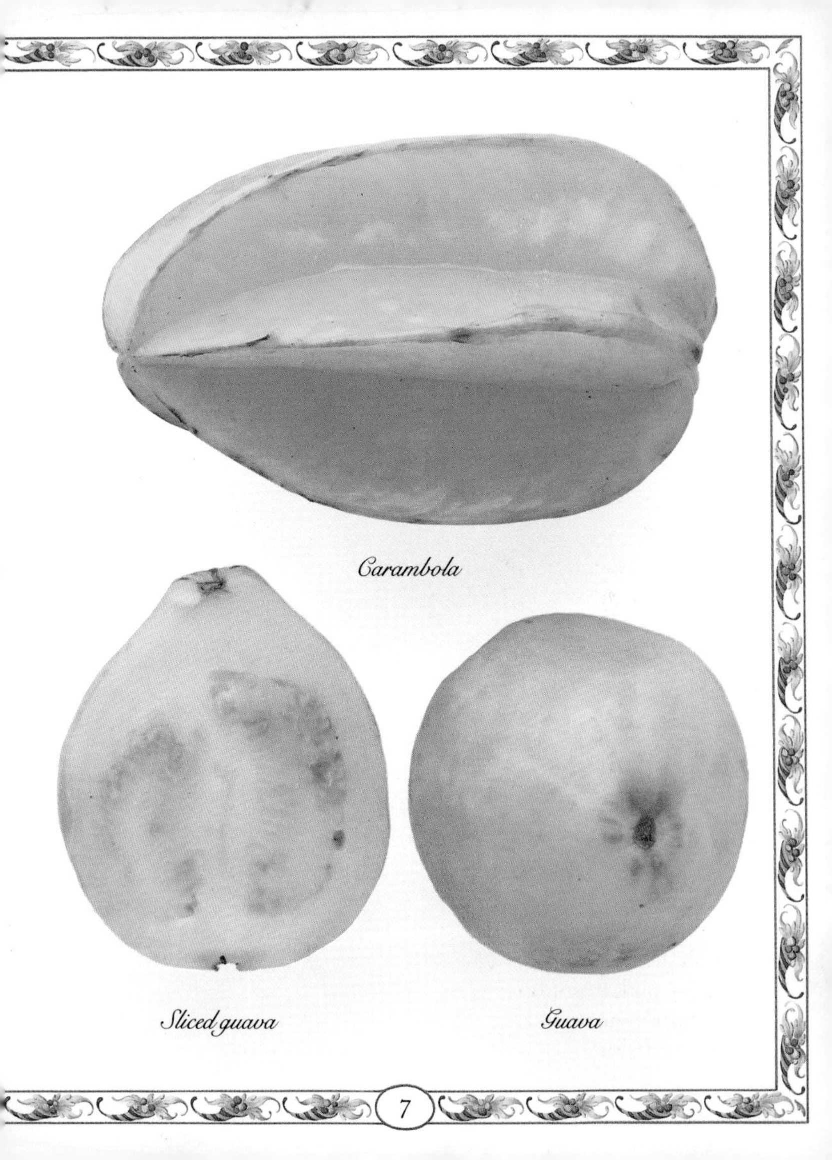
Carambola

Sliced guava

Guava

FIG

FIGS, Ficus carica, *are a fruit of antiquity. With wheat, olives and grapes they were one of the staples of the Mediterranean diet. Native to Asia Minor, figs now grow worldwide. Round or pear-shaped, they have tender, thin skin which may be green, brown or purple; the flesh is white or white and red, depending on the variety.*

Fresh figs

Ripe figs are soft, juicy and taste slightly nutty. They are best straight from the tree. Good quality semi-dried and dried figs make a good chewy snack or a compote.

POMEGRANATE

THE POMEGRANATE, Punica granatum, *is another ancient fruit long naturalized throughout the Middle East and Mediterranean region. Recorded in the* Book of Moses *together with figs and grapes, and buried in Egyptian tombs, pomegranates were thought holy or mystical by many civilizations, perhaps because of their appearance. The thick, red or brown, leathery skin hides scarlet, translucent beads set in a white membrane.*

Pomegranates have a distinctive sweet-sour taste. The seeds are good in salads, cold soups and desserts. The juice makes a refreshing drink and the sourish syrup sold in Middle Eastern shops makes unusual sauces for lamb and poultry.

Pomegranate

Pomegranate seeds and juice

ASIAN PEAR

ASIAN PEARS, Pyrus sinensis, *are found under a variety of names: oriental pear, Tientsin pear, sand pear,* nashi *(Japanese for pear). They are apple-like in shape and range in colour from pale yellow to russet brown. The flesh is refreshing and mild and has lots of juice. They do not become soft like European pears; the texture remains crunchy even when ripe. Good in vegetable and fruit salads, but if cooked they need longer than Western pears.*

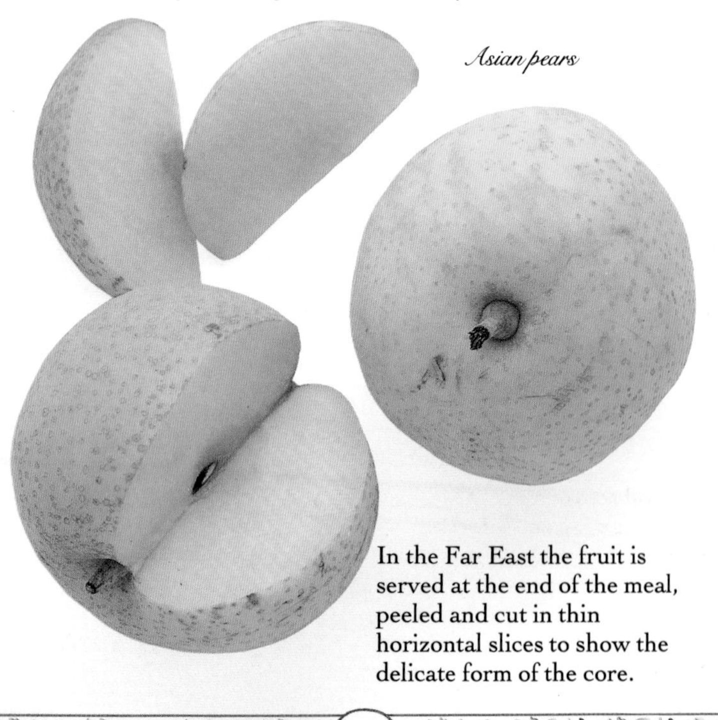

Asian pears

In the Far East the fruit is served at the end of the meal, peeled and cut in thin horizontal slices to show the delicate form of the core.

TAMARILLO & PHYSALIS

Physalis, Physalis peruviana, *and tamarillo,* Cyphomandra betacea, *belong to the tomato family and are native to Central and South America.*

Physalis, or Cape gooseberries, resemble cherry tomatoes wrapped in a parchment-like calyx. The fruit may be golden, red or purple, and keeps well in the husk. The sharp-sweet flavour is part tomato, part grape, part gooseberry.

The oval tamarillo, or tree tomato, is deep yellow or red with golden flesh and purple seeds. The yellow variety is sweeter, but neither seems ripe enough to eat raw in Europe. Always remove the shiny, bitter skin.

QUINCE & MEDLAR

QUINCES, Cydonia vulgaris, *native to western Asia, are now cultivated worldwide. Apple- or pear-shaped, some covered with a soft ash-coloured down, quinces were perhaps the 'golden apples' of legend. The ripe fruit has an intense, warm aroma that fills the room; the flavour is fragrant, well complemented by sweet spices.*

Quince

Quince segment

Quinces must be cooked before eating. In the Middle East they are stuffed with meat mixtures and used in stews. Quince paste from France and Spain is an excellent thick preserve.

Seldom seen in commerce, medlars, *Mespilus germanica*, look like russet apples with a large calyx. Keep until overripe or they are not good to eat.

Medlar tree branch

LOQUAT

A SUBTROPICAL FRUIT *long esteemed in China and Japan, loquats,* Eriobotrya japonica, *now grow widely in the Americas and around the Mediterranean. The apricot-coloured fruit has a downy skin and a firm, pleasant texture. It ripens early in spring, and its delicate flavour is clean and refreshing.*

Loquats

Unfortunately loquats do not travel well. They bruise easily, so most of the fruit is picked and shipped when underripe. In this condition they are disappointing; loquats should ripen on the tree until they are tender and scented. Tart loquats are best poached in syrup or made into jam (they have a high pectin content).

DURIAN

THE DURIAN, Durio zibethinus, *is a controversial fruit. In tropical Asia it is the king of fruit; to Westerners its fetid and garlicky smell overwhelms and stops them trying it. Discard the strong-smelling skin, chill the creamy pulp and eat outdoors. The flavour is sweet yet piquant, and durian is rich in protein and carbohydrate.*

Durian

CHERIMOYA

Cherimoyas belong to the Anona *family of which there are some 60 different species. Native to the South American tropics, they are now widely cultivated in Spain, Israel, New Zealand and Australia. The creamy flesh covers large black seeds and is delicate and easy to digest. It tastes subtly of banana, vanilla and mango.*

Cherimoyas, or custard apples, *A. cherimola*, look like closed green pine cones and can be the size of an apple or as big as a melon. Sugar apples (*A. squamosa*) have a less fine taste. Bullock's heart (*A. reticulata*) has more solid flesh and is red-brown.

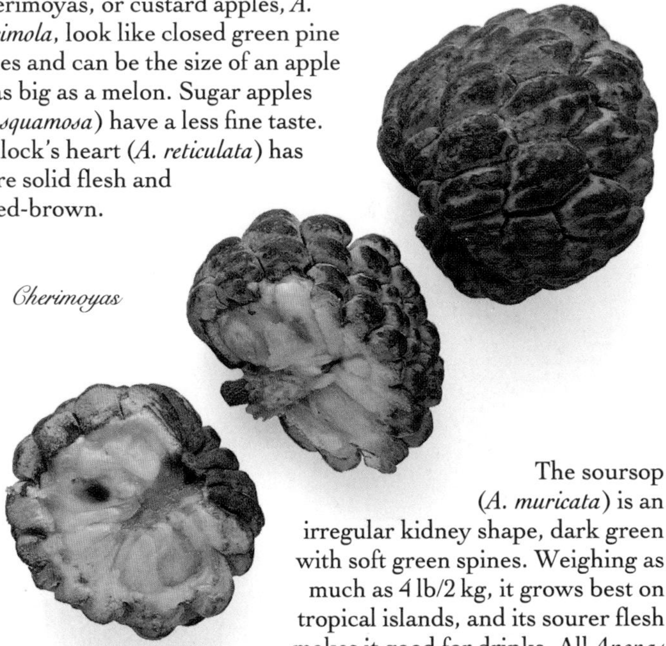

Cherimoyas

The soursop (*A. muricata*) is an irregular kidney shape, dark green with soft green spines. Weighing as much as 4 lb/2 kg, it grows best on tropical islands, and its sourer flesh makes it good for drinks. All *Anonas* are sensitive to extremes of heat and cold, so store at room temperature.

PAPAYA, BABACO & MANGO

Papayas, or pawpaws, Carica papaya, *are one of the most common tropical fruits, now planted worldwide. When Columbus discovered them in the Caribbean islands he called them the fruits of angels. The fruits hang in clusters like coconuts on a tall plant; they can weigh up to 10 lb/5 kg. The skin is inedible; the flesh perfumed and reminiscent of apricot. Great for breakfast with a little lime juice.*

Papaya

Papaya leaves and fruit contain a protein-tenderizing enzyme, papain, which is a good aid to digestion. Babaco, *C. pentagona*, is the mountain papaya. Five-sided, it may be long and pale green or golden and quince-like. It has a clean, fresh taste.

A good mango is the finest of the tropical fruits. Cultivated in India for 4,000 years, *Mangifera indica* from the wild gives small fibrous fruit smelling of turpentine.

Sliced mango

Mango

Cultivated mangoes have a fine texture, a good acid/sweetness balance and a flavour of peach, melon and spice. Colour does not show ripeness: they may be green, red or orange-pink; a ripe fruit yields to gentle pressure. Ripe mango is messy but delicious eaten just as it is.

Babaco

Babaco slice

KUMQUAT, POMELO & UGLI FRUIT

CITRUS FRUITS *originated in eastern Asia in the tropical and subtropical zones. Today the citrus belt covers the world and the original handful of species has been selected, crossed and re-crossed so many times to produce dozens of varieties, with much confusion about names and relationships.*

Kumquats

Ugli fruit

Pomelos

Kumquats
'are eaten without
peeling their golden coats.
When preserved in honey,
their flavour is still better'.
Han Yen-Chih, 1178

Pomelos, or shaddocks, are smooth, pale yellow-green with a thick rind and bitter pith that comes off easily when quartered. The pink or yellow flesh separates cleanly into segments and has a light, sweetish taste and little juice.

Ugli fruit is a cross of grapefruit, Seville orange and tangerine, developed in Jamaica. The misshapen, lumpy fruit has few seeds and a fragrant, juicy flesh, sweeter than grapefruit.

PRICKLY PEAR

PRICKLY PEARS *grow in clusters on the cactus of the same name. They are orange or red when ripe, and the flesh inside may be red or yellow with a pleasant, bland taste to which the small seeds add a crunchy texture. Native to Central America,* Opuntia ficus-indica *has now spread to many hot, dry regions. The fruit is refreshing to eat on a hot day.*

Wear rubber gloves to peel the fruit: slice off the ends, make a vertical slit, not too deep, and ease the knife between the skin and flesh on either side. Flatten the skin and lift off the barrel-shaped fruit.

Prickly pear

PERSIMMON

*I*N WINTER, BRILLIANT *red persimmons hang on the leafless trees of Provence like small lanterns. The persimmons we have,* Diospyros kaki, *come from Japan where the fruit is much prized. Unripe persimmons are inedible; before eating, they must be kept until almost rotting.*

Sharon fruit

Persimmons

Slice the top off a ripe persimmon, scoop out the flesh and eat it or mix with a little cream. Sharon fruit, developed in Israel, has had the astringency and the seeds bred out of it. It can be eaten while firm; it goes well in salads and with soft white cheeses. By comparison with persimmon the taste is bland and dull.

PASSION FRUIT & GRANADILLA

*P*ASSION FRUIT *and granadilla belong to the same family as the blue flowering plants that look so striking in conservatories and warm gardens.* Passiflora edulis *is native to Brazil, but now more widely planted throughout the tropics.*

Ripe passion fruit looks rather battered but it has an intense, fragrant aroma of lime, musk and jasmine, which remains even in ice cream or mixed with other fruits. The juice makes a perfect summer drink. The smooth, orange globes of granadilla, so light in the hand, look stunning but taste less fine.

Passion fruit

Granadillas

LYCHEE & RAMBUTAN

LYCHEES AND RAMBUTANS *are related, but lychees,* Litchi chinensis, *come from China and rambutans,* Nephelium lappaceum, *from Malaysia. For 2,000 years lychees have been regarded as the finest fruit in China, once rushed to the court by relays of horsemen, and required as payment for taxes in some districts. The tall tree has evergreen, coppery leaves with the deep pink or buff fruits hanging in clusters.*

Lychees

Rambutans

Rambutans look rather strange with their red skins and soft, curved spines which feel a bit like blades of grass. The skin is thin and pliable, the fruit inside, around an almond-like stone, is very similar to the lychee.

Inside the rough, warty skin, lychees have a translucent white fruit, rather like a grape in texture. It is very juicy and tangy but sweet, with a perfume of cherries and roses. Both fruits are high in vitamin C.

MANGOSTEEN

NATIVE TO MALAYSIA, *this striking fruit is now grown more widely in Southeast Asia and in South America. The tough, purple skin reveals a deep pink pith, which rapidly browns when cut, and a waxy, white fruit. The lobes in the calyx on the underside usually equal in number the segments in the fruit.*

Whole mangosteen

Mangosteen segments

Mangosteens, *Garcinia mangostana*, have an indefinable, delicate aroma that is reminiscent of apricots, grapes and pineapple. The flesh is soft and refreshingly balanced in acidity and sweetness. Mangosteens are rightly called the queen of tropical fruits. Picked when fully ripe, they are often past their best by the time they arrive in Europe.

SAPODILLA

SAPODILLA, Manilkara zapota, *came originally from Central America, but has been more widely cultivated across the tropics for several centuries. The sapodilla tree is best known for its sap, the source of chicle, which is used to make chewing gum.*

Halved sapodilla

The fruit looks a bit like a potato with its dull brown skin. To determine ripeness, scratch the skin: if the scratch is green the fruit is unripe, if yellow, ripe. Sapodillas are picked unripe; they travel and keep well, but should be almost overripe when eaten, otherwise they are too astringent. The grainy flesh tastes of floral honey and peach. Eat as a dessert fruit.

Whole sapodilla

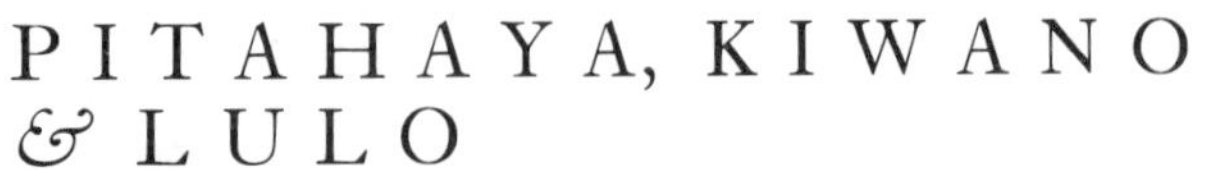

PITAHAYA, KIWANO & LULO

Pitahayas, which look somewhat like small, bumpy pineapples or pine cones, are a cactus fruit from Colombia. The light-grey translucent flesh with its crunchy seeds is sweet and bland. It needs a squeeze of lime juice to sharpen it. Eat from the skin with a spoon, or add to fruit salads.

Whole pitahaya

Lulo

Sliced pitahaya

Kiwanos, or horned melons, are native to tropical Africa but now come from California and New Zealand. The vivid orange skin flamboyantly reveals emerald green flesh, but aroma and taste are lacking, and the jelly-like texture combined with the large seeds is disagreeable. Maybe this is because the fruit we get is not sufficiently ripe.

Lulos are wild fruits gathered in the rain forests of South America. When ripe the fleshy skin is golden-yellow, but the pulp and seeds have a sharply acid taste in Europe – perhaps they need to ripen on the tree.

Halved kiwano

Whole kiwano

Halved lulo

All the recipes are for 4
but some (such as preserves)
will serve more

PAPAYA AND SERRANO HAM

Peel *1 or 2 papayas*, depending on size, and cut them in thick slices. Sprinkle with *lime juice* if you wish and arrange on a serving platter with *8 oz/250 g serrano ham*.

Variation

If you cannot get serrano ham use another well-flavoured cured ham, such as Bayonne.

PEAR AND WATERCRESS SALAD

2 Asian pears
1 bunch watercress
2 tablespoons hazelnuts
2 tablespoons hazelnut or walnut oil
2 tablespoons olive oil
2 tablespoons white wine vinegar
salt and pepper

Peel and slice the pears. Remove any long stalks from the watercress. Toast the hazelnuts in a preheated oven, 180°C/350°F/gas 4, for 10 minutes, then rub off their skins in a tea towel and chop coarsely.

Make a dressing with the remaining ingredients. Put the pears, cress and nuts in a salad bowl, spoon over the dressing and serve.

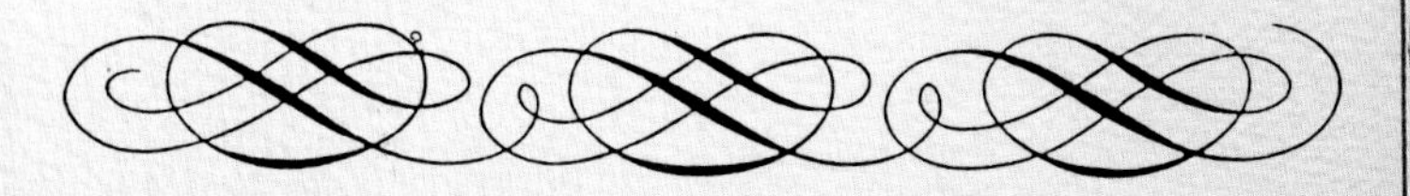

RED KIDNEY BEANS WITH WALNUTS AND POMEGRANATE

A Georgian dish that is served cold as a first course.

3 oz/75 g walnuts
1 clove garlic
salt
5 tablespoons pomegranate juice
cayenne
1/4 teaspoon allspice
1/2 small onion, chopped finely
2 tablespoons chopped parsley
2 tablespoons chopped mint
8 oz/250 g red beans, cooked
seeds from 1/2 pomegranate

Put the walnuts in a food processor and grind. Crush the garlic with a little salt and add, together with the pomegranate juice. Blend to a thin paste, adding a little more juice, or water, if necessary. Season with cayenne and allspice, and salt if necessary. Mix the paste, onion and most of the herbs gently into the beans, and garnish with the pomegranate seeds and remaining herbs.

PAPAYA AND SEAFOOD SALAD

6 oz/175 g young spinach leaves
1–2 papayas, depending on size
½ cucumber
a handful of pine nuts
juice of 2 limes
1 tablespoon soft brown sugar
*2 tablespoons fish sauce**
a small piece of fresh ginger, chopped finely
2 tablespoons chopped coriander leaves
2–3 spring onions, chopped
1 green chilli, seeded and chopped
8 oz/250 g medium prawns, cooked
8 oz/250 g white crab meat, shredded

Shred the spinach leaves. Cut the papaya and cucumber into cubes. Toast the pine nuts in a dry frying pan until golden. Combine the lime juice, sugar, fish sauce, ginger and coriander to make a dressing. Arrange the spinach in a salad bowl; add the other ingredients; spoon over the dressing and serve.

*Fish sauce is a clear brown liquid that is drained off fish that has been fermented in brine. It is available from oriental shops.

FENNEL, AVOCADO AND KUMQUAT SALAD

1–2 heads fennel
1–2 avocados
10 kumquats
4 tablespoons olive oil
3 tablespoons white wine vinegar
salt and pepper

Remove the stringy outside layers from the fennel and cut the inner flesh into thin strips. Slice the avocados. Wash the kumquats, slice them thinly and pick out any pips. Arrange all the ingredients in a salad bowl. Make a dressing with the other ingredients, spoon over the salad and serve.

HALIBUT STEAKS WITH CARAMBOLA

A good recipe for using unripe carambolas. Other firm white fish steaks or fillets could be used instead of halibut.

1 oz/25 g butter
4 halibut steaks
a small piece of fresh ginger, chopped finely
1 teaspoon paprika
salt
lemon juice
2 carambolas

Butter an oven dish into which the steaks will fit in one layer. Rub the steaks on both sides with ginger, paprika and salt. Put them in the dish and sprinkle with lemon juice. Top and tail the carambolas and cut in thin slices. Spread the slices over the fish. Dot with the remaining butter and cover with foil. Bake in a preheated oven, 200°C/400°F/gas 6, for 15–20 minutes, depending on the thickness of the steaks.

LAMB AND QUINCE TAGINE

A tagine is a Moroccan stew, usually made with lamb or chicken and a variety of spices, vegetables or fruit.

2 lb/1 kg lean lamb, cubed
2 large onions, chopped
2 oz/50 g butter
1/2 teaspoon ground ginger
a few saffron threads, crushed
1/2 teaspoon freshly ground pepper
1 lb/500 g quinces
salt

Put the lamb, 1 onion, the butter and spices into a heavy pan and add water just to cover. Simmer the tagine, covered, for an hour or more, until the meat is very tender. Wash the quinces, cut in half or in four, according to their size, remove the core, but don't peel them.

Add the remaining onion and the quinces to the pan with salt to taste, and simmer until the quinces soften. This can take anything from 15 to 30 minutes; make sure they don't fall apart. If there is still a lot of liquid in the pan, lift out the meat and quinces and turn up the heat to reduce it; if not, serve at once.

MANGO PILAF

8 oz/250 g basmati rice
salt
3/4 pint/450 ml water
1 large mango, not too ripe
3 oz/75 g butter
1/4 teaspoon ground cardamom
1 teaspoon turmeric
1/4 teaspoon ground nutmeg
1/4 teaspoon ground ginger
a handful of cashew nuts

Wash the rice. Add salt to the water and bring to the boil. Put in the rice, bring the water back to the boil, and simmer, covered, for 12–15 minutes, until the water has been absorbed and the surface of the rice is covered with holes. Peel the mango and cut the flesh into cubes. Melt the butter in a frying pan and fry the spices for 2 minutes, stirring frequently. Add the pieces of mango and cook for a minute or two more, very gently so that the fruit doesn't break up. Toast the cashew nuts in a preheated oven, 180°C/350°F/gas 4, for 10 minutes.

Turn the rice out with a fork into a warm bowl. Stir in the mango, spices and butter and scatter the cashews on top.

CHERIMOYA ICE CREAM

2 ripe cherimoyas
1 oz/25 g caster sugar
3 tablespoons Cointreau or lime juice
1/4 pint/150 ml double cream

Scoop the flesh from the cherimoyas, discard the pips, and purée it in a food processor or blender with all the other ingredients. Freeze in an ice cream machine for about 10 minutes, or put in a shallow plastic box in the freezer until the mixture has set around the sides of the box. Then tip into a chilled bowl, beat well and return to the box and to the freezer.

GUAVA CREAM

6 ripe guavas
1–2 tablespoons caster sugar
1–2 tablespoons kirsch
1/4 pint/150 ml cream

Peel the guavas and chop them into fairly small pieces. Whisk the sugar and kirsch into the cream – let the amount depend on the sweetness of the guavas. Stir in the fruit and chill for an hour or more before serving.

Variation

Feijoas can be prepared in the same way.

MANGOSTEEN AND STRAWBERRY SALAD

Remove the thick skins from *8–10 mangosteens* and divide the fruit into segments. Put them into a bowl with *8 oz/250 g strawberries* and *4 tablespoons dry white wine*. Leave to stand for 30 minutes or more before serving.

SPICED UGLI FRUIT AND POMELO WITH RUM

You can make this refreshing citrus dessert with other fruit combinations.

2 ugli fruits
1 pomelo
2 tablespoons brown sugar
1 teaspoon cinnamon
a pinch of ground cloves
2–3 tablespoons rum

Peel the ugli fruits, removing all the pith. The pomelo has a thick skin, so cut the skin and pith lengthways in quarters and pull off. Keep to make candied peel (see p. 37). With a sharp knife slice along the membrane partitions to free the fruit segments. Work over a bowl to drop the fruit into and to catch the juice. Taste and add sugar as necessary. Then sprinkle with the spices and the rum and mix well. Leave to stand for 30 minutes before serving.

LOQUATS IN SYRUP

½ pint/300 ml water
5 oz/150 g sugar
peel from 1 lemon
1 tablespoon whole cloves
1 lb/500 g loquats

Bring the water, sugar, lemon peel and cloves slowly to the boil, and simmer for 3–4 minutes. Pick off the flower ends from the loquats and remove the skins – they come off quite easily. Add the fruit to the syrup, and poach gently for about 5 minutes, until the loquats are soft. Leave to cool in the syrup.

CLAFOUTIS WITH LYCHEES

3 eggs
½ pint/300 ml milk
*1 oz/25 g vanilla sugar**
a pinch of salt
3 oz/75 g plain flour, sifted
12 oz/375 g lychees, peeled and stoned

Put the eggs, milk, sugar, salt and flour into a liquidizer and blend. Lightly butter a shallow 8 in/20 cm baking dish or flan tin and pour in a little of the mixture. Spread the lychees over the batter and pour the rest around them. Bake in a preheated oven, 200°C/400°F/gas 6, for about 30 minutes, until the clafoutis is puffed and browned.

*Vanilla sugar is made by keeping a vanilla pod in a jar of caster sugar, and topping up the sugar as you use it.

FIGS WITH RASPBERRIES

8 black figs
8 fl oz/250 ml double cream
2 tablespoons eau de vie de framboise
1 tablespoon caster sugar
4 oz/125 g raspberries

Peel the figs and cut into quarters. Whip the cream lightly with the eau de vie and sugar. Arrange the fruit on dessert plates and serve with the cream.

FIGS IN SYRUP

A good sweetmeat to serve with coffee after dinner.

2 lb/1 kg small figs
blanched almonds
juice of 1/2 lemon
1 1/2 lb/750 g sugar
3/4 pint/450 ml water
4 cloves

Wipe the figs and cut off the stalks. Cut a small slit in the stalk end of each one and push in an almond.
Bring the lemon juice, sugar and water to the boil to make a syrup. Simmer for a few minutes, then add the figs and the cloves. Remove from the heat and leave to steep overnight.
The next day, bring slowly to the boil and simmer until the figs are just soft. Lift them out into a large jar. Boil down the syrup a little if necessary, then pour over the figs. Leave to cool and then close the jar.

PERSIMMON AND PINEAPPLE ICE CREAM

3 large ripe persimmons
2 slices fresh pineapple
2 oz/50 g caster sugar
3 tablespoons white rum
1/2 pint/300 ml double cream

Make sure the persimmons are very ripe or they will taste harsh and acidic. Scoop out the pulp and sieve. Cut the pineapple into small pieces. Purée the fruit in a processor with the sugar and rum. Whisk the cream lightly, stir it into the fruit and freeze in an ice cream machine for about 15 minutes, or follow the instructions on p. 33.

CANDIED POMELO PEEL

An unusual accompaniment to after-dinner coffee, candied pomelo peel has a clean and slightly bitter citrus flavour.

peel from 1 pomelo
1 1/4 lb/625 g sugar
8 fl oz/250 ml water

Remove all the membrane from the peel, but make sure the pith is still attached to it. Cut it into long strips about 1/4 in/5 mm wide. Put the peel in a pan, cover with water, bring to the boil and simmer for 5 minutes. Drain and repeat the process twice more with fresh water. Make a syrup with 1 lb/500 g of the sugar and the water in a heavy pan. Boil for 1 minute, then add the peel. Cook slowly for 30–40 minutes, until the peel is soft. There should be very little syrup left in the pan. Turn the peel out onto a sheet of foil or waxed paper and leave to cool, then sprinkle with the remaining sugar. Leave to stand, uncovered, until the peel is completely dry – it can take several days – then store in a jar. It will keep almost indefinitely.

TAMARILLO CHUTNEY

1 tablespoon oil
6 shallots, peeled and chopped
4 cloves garlic, peeled and chopped
2 tamarillos
1 red chilli, seeded and chopped
1 teaspoon ground allspice
1/2 teaspoon ground mace
1/4 teaspoon ground cloves
seeds from 4 cardamoms
1 1/2 tablespoons brown sugar
1 green apple, chopped
3 tablespoons wine vinegar
salt

Heat the oil and cook the shallots and garlic over low heat for a few minutes, until soft.

Pour boiling water over the tamarillos to loosen the skins, then peel and chop them. Add to the pan with all the other ingredients, and cook gently for about 20 minutes, stirring frequently, until the chutney becomes thick and jam-like.

This quantity will fill a 1 lb/500 g jar.

FONDANT-COATED PHYSALIS

A plate of fondant-coated physalis makes a good end to a meal instead of a dessert. Eat them on the day you make them.

12 physalis
4 oz/125 g fondant

Choose ripe fruit and peel back the papery covering, cutting it in one or two places if necessary. The fruit must be dry for dipping; fondant will not set on a moist surface.

Put the fondant in a small bowl and melt it slowly over hot water, stirring all the time. If it is too thick add a tablespoon or two of water. Holding the fruits by their papery wings, dip into the fondant, one or two at a time. Let the excess drip off, and hold each fruit until it dries, then leave on a sheet of waxed paper for 5–10 minutes until the fondant has hardened properly. Serve in small paper cases.

*C*ARAMBOLAS IN HONEY

This is an easy, rich-tasting preserve.

Slice *2 or 3 ripe carambolas* crossways into a bowl. Cover with *honey* and leave to stand overnight. The next day, bring gently to the boil with *1/2 stick of cinnamon*, simmer for 4–5 minutes, then spoon the fruit into jars and pour over the honey.

*M*ANGO LASSI

A refreshing Indian yogurt drink that is simple to make. For 2 servings take:

1–2 ripe mangoes
1/2 pint/300 ml plain yogurt
2 tablespoons honey (optional)
4 fl oz/125 ml water

Cut the mango in half around the large flat stone and scoop out all the flesh. Put it into a blender with all the remaining ingredients and whizz until the drink is frothy. Chill and serve.

*P*ASSION FRUIT PUNCH

12 passion fruit
1/2 pint/300 ml water
a squeeze of lemon juice
juice of 2 oranges

Cut the tops off the passion fruit and scoop the pulp into a sieve. Scrape through as much pulp and juice as possible. Add the other ingredients and chill.

Variation

Add 2 measures of light rum or vodka.

INDEX

ACKNOWLEDGEMENTS

The publishers would like to thank the following:

JACKET
· PHOTOGRAPHY ·
DAVE KING

· TYPESETTING ·
TRADESPOOLS LTD
FROME

· ILLUSTRATOR ·
JANE THOMSON

PHOTOGRAPHIC
· ASSISTANCE ·
JONATHAN BUCKLEY

· REPRODUCTION ·
COLOURSCAN
SINGAPORE

PAGES **5**, **34**, MARY EVANS PICTURE LIBRARY, LONDON
MARIA JOSE SEVILLA AND FOODS FROM SPAIN

ROSIE FORD FOR ADDITIONAL HELP